the guide to owning
Cichlids

Richard F. Stratton

Dedication

To Ashley

T.F.H. Publications, Inc.
One TFH Plaza
Third and Union Avenues
Neptune City, NJ 07753

Printed and bound in China
06 07 08 09   3 5 7 9 8 6 4

This book has been published with the intent to provide accurate and authoritative information in regard to the subject matter within. While every precaution has been taken in preparation of this book, the publisher and author assume no responsibility for errors or omissions. Neither is any liability assumed for damages resulting from the use of the information herein.

ISBN 0-7938-0371-3

www.tfh.com

Contents

Rainbow Cichlids are small, golden-colored fish from Central America. They are territorial but not overly aggressive.

Cichlid Envy

Cichlids are so popular that there are probably a lot of people keeping them, or trying to, that should not. That may be a heretical statement to those who wish to stoke the fires of the nearly unreal cichlid mania that has swept the aquarium world for the last three decades, but the point is that most cichlids are not a fish for beginners. Naturally, neophytes can't be prohibited from keeping these very interesting animals, but they should at least be warned that there can be problems. The problems are that cichlids are generally territorial and aggressive. This means that they fight a lot, and they can quickly kill tankmates, fellow cichlids or not. The aggression is not out of meanness; it has served the family well in its spectacular evolutionary success in their efforts to protect their eggs, their young, and their territory. Still, it is a problem to be managed in the aquarium, and one of the problems is that the very fact of confinement in an aquarium can exacerbate the aggression of these animals.

Another problem with cichlids is that many species get quite a bit larger than the average aquarium fish. That means that they have more of an impact on the water, so special filtration and partial water changing regimens will be needed. Obviously, large aquaria are needed, but I am often amazed to recall how we made do with small tanks before the advent of Space Age technology and the discovery of new sealants and other materials that made large tanks practical and inexpensive to manufacture. Early cichlid hobbyists pioneered the regular partial water changes that are so often recommended today for all tropical fish hobbyists, only for cichlidophiles (cichlid lovers) the "regular partial water changes" were an everyday thing instead of every two weeks. This was especially the case when the cichlids in question were crowded into tanks that were too small for them. Those hardy pioneer cichlid specialists also made use of refrigerator liners and just about everything but horse troughs in an effort to keep their ever-expanding collection of constantly growing fishes.

Since so many new hobbyists become enamored with cichlids, it is my intention to cover the most basic elements of cichlid husbandry, along with a few secrets of the trade that are not always shared outside the cichlid fraternity of enlightened men and women (and a few precocious children). I also plan to present an overview of the species in different parts of the world, with recommendations for some types to begin with from each region. In the meantime, it is worth considering what all the shouting is about. Why are cichlids so popular with everyone, newcomer and experienced hobbyist alike? Why do those who don't have cichlids envy you for keeping them?

WHY KEEP CICHLIDS?

One of the primary attractions of cichlids is that of all the fishes, they are the masters of parental care. Every single cichlid species provides some type of care not just for the eggs but for the fry once they have hatched. This is not only in marked contrast to most other fish species, but the care and protection cichlids invest in their young exceed that of amphibians and reptiles, not to mention some species of birds and mammals. Cichlids offer many other attractions, but it is this tender devotion lavished upon the young that attracts so many hobbyists to this remarkable family. Ironically, the same behaviors that touch our hearts also tend to vex our souls. The same aggression that elicits our admiration when a tiny cichlid parent attacks the hand of the "huge monster" (the keeper) who is obviously menacing its young draws a different response when it is directed toward a favorite and expensive fish!

Scientists also are attracted toward cichlids for various reasons. The first is that cichlids have many species that are quite interesting to study in regard to their behavior. Since in many cichlid species the male and female cooperate in the care of the young, it gives ethologists (scientists who study behavior in animals) an opportunity to study communication between individual fish and to determine how they recognize one another. The last part would be of particular interest to any person who has ever seen a guarding pair of parental cichlids. Not content to merely guard the young, the pair takes turns making forays throughout the tank, looking for enemies. A typical setup to show this behavior would be a cichlid community tank with about six Firemouths, some Convict and Rainbow Cichlids, and a non-cichlid "dither fish" such as a fast-swimming barb to take some of the aggression. I have many times seen one of a pair of Firemouth Cichlids returning from a foray to join the guarding partner. To me the fish may be identical, but the guarding Firemouth is immediately appeased upon

Firemouth Cichlids make marvelous parents and are quite ferocious in defense of their fry.

viewing her mate; she recognizes him in some visual way that is instantaneous. In at least some species, such as jewel cichlids, scientists have evidence that the iridescent vermiform (worm-like) markings on the head are important in one individual recognizing another, but Firemouths don't have those markings, so there is much to learn.

While we may not always be able to tell individual specimens of some cichlid species from one another, some of them are able to recognize us. Not only does a tank of cichlids learn to distinguish which member of the family feeds them, but there are more remarkable examples of cichlid discrimination. Occasionally one of the larger cichlids is kept by itself in a large tank as a sort of family pet. When one of the large cichlids is kept alone in this manner, the individual tends to interact more with humans.

Do Jewel Cichlids use the iridescent markings on the heads of their kin to recognize individuals?

THE GUIDE TO OWNING CICHLIDS

Another reason that scientists study cichlids is to try to ascertain the reason for their phenomenal success. Some scientists believe that cichlids are the most successful family of all vertebrate species. This is based on the fact that cichlids have evolved into some 1400 species (a conservative estimate, as many species have not been scientifically described yet) and dominate the fish fauna in many areas. The Great Lakes of Africa are veritable fish bowls of cichlid species, with many of them being endemic (found only there) to that particular lake. Cichlids are adaptable, but the question has always been: Why is this so? The answer is not certain, but some scientists hazard the suggestion that cichlid intelligence may be part of the reason.

The success of cichlids biologically is a part of their success with tropical fish hobbyists. Everyone loves a winner, and certainly the family Cichlidae seems to be one. But not everyone is happy about it, as cichlids that have been introduced by accident or by design (sometimes as gamefishes) compete to the disadvantage of native fish faunas around the world. In the US, the most infamous introductions have been in Florida, as the one thing that cichlids have not evolved to tolerate is cold water. The only naturally occurring cichlid in the US. is the Texas Cichlid, which is found in the southern reaches of Texas and in northern Mexico.

There are many different cichlids available, and many of them are quite beautiful in form and of dramatic coloration. This fact has added to the popularity of the family, as everyone loves a tank of beautiful fish. Additionally, cichlids are generally a hardy lot. Some species seem to be so tough and adaptable that a friend used to comment that "You couldn't kill one with a hammer." Of course, this is only true of cichlids in general, as there are many species that could be described as delicate and difficult to keep. Such species simply add to the popularity of the family, as there are always those aquarists who like nothing more than a challenge.

SIZE

The only thing that the cichlid family has working against it as an aquarium fish is that many species are quite large, and the general run of cichlids is a little on the large side for aquarium fishes. This apparently is not a serious problem, as it is precisely the larger species that are quite popular. For those who want smaller fishes, there are many species of "dwarf" cichlids. Such species can be gentle, like the ever-popular Ram, or they can be replicas of their fire breathing larger brethren on a smaller scale.

Since cichlids come in all sizes, have interesting behavior, are generally hardy, and are something of an enigma to scientists, it is not surprising that the family has been so popular with both aquarists and scientists. Cichlids not only have a special appeal, but they create envy in our friends who can't have them. What more could we want?

Secrets of Cichlid Success

In order to understand just how successful cichlids, family Cichlidae, have been, it is necessary to understand their distribution and some of the niches in which they have evolved. The more than 1400 cichlid species range from the disk-like angelfishes (*Pterophyllum*) and discus (*Symphysodon*) to tackle-busting Wolf Cichlids (*"Cichlasoma" dovii*), not to mention tiny shell-dwellers in Lake Tanganyika and diminutive *Apistogramma* species in South America. All of these

The special shape of the head of the Calvus Cichlid enables it hunt in the rocky crevices in its natural environment.

Neolamprologus ocellatus **is one of the tiny shell-dwellers of Lake Tanganyika.**

species make their living exploiting a vast array of food resources, from algae to higher plants and from minute crustaceans to crabs and large fishes. Cichlids tend to be lurking predators, but some are built like barracudas and chase down their prey by sheer swimming speed. Others, such as *Lamprologus (Altolamprologus) calvus*, have a compressed shape to fit into crevices in rocks to attain hiding crustaceans and small fishes. Some of the pike cichlids (*Crenicichla*) are downright toad-like in their ability to hide in the mud and pounce on prey. At the other end of the spectrum, schools of *Tropheus* species in Lake Tanganyika descend upon algae-laden rocks to consume that resource. Such fishes are hardly mindless herbivores, as they have feeding territories, something that is rare among freshwater fish species, and they have a complex social order, which has made them difficult to keep in captivity.

Although cichlids are quite modern fishes, it is believed that they all descended from an ancient marine ancestor. Such early roots would help explain the distribution of the family, since there is good evidence that South America and Africa were one continent before being separated by the drifting of tectonic plates (continental drift).

Cichlids are intelligent animals for fishes; in effect, they have good software to go with some very good hardware. They have behaviors that include some adaptability, and perhaps that is one

Agassiz's Dwarf Cichlid, an *Apistogramma*, is a native of the Amazon and its southern tributaries. The species is distinguished by the colorful extended wedge-shaped tail.

reason why many of them spawn well in our aquaria. Cichlids have a certain plasticity because they have a lot of variation in their genetic makeup. Although this factor is dismissed as circular reasoning by many scientists, there is no doubt that generalized fishes such as cichlids can more easily evolve to fill new niches than can highly specialized ones. There is some controversy over this concept, but there is ample evidence for variation or plasticity being a distinct advantage with cichlids.

CICHLID BIOLOGY

One advantage that cichlids share with wrasses, pomacentrids, parrotfishes, and surf perches is that they have well-developed pharyngeal teeth in the throat in addition to more normal dentition in the mouth. This distinction has inspired some scientists into grouping these families together as a suborder, the Labroidei. That is not to say that there are not other groups of fishes that have teeth both in the throat (pharynx) and the jaws, but it is very uncommon, and in most cases the teeth are only feebly developed. In these labroid families the teeth and the jaws are highly evolved to work together in different ways. This is one of the advantages to having the two sets of teeth: the jaw teeth can be specialized for one activity, while the pharyngeal teeth can be specialized for yet another. For example, a predator may have sharp canine-like teeth in the jaws for capturing other fish, while the teeth in the pharynx have the purpose of grinding up the prey once it has been captured and is in the process of being engulfed. The advantages of this system in evolving to fill different niches are clear. The jaw and pharyngeal teeth can evolve in different directions to perform different functions and to supplement one another.

It is thought that the cichlid jaw is exceptionally structured to be able to assume a variety of forms to fulfill numerous different applications. Unlike primitive fishes, cichlids have a bone (the premaxilla, often paired and then called premaxillae) at the tip of the upper jaw that is not immovably bound to the cranium but is free to move. The ability of this jaw bone to slide on a pedicel or stalk enables a wide variation of mouth forms to evolve, from almost tube-like structures for inhaling small prey to heavy anvil-like jaws for seizing larger prey to solid chisel-like jaws for scraping algae off rocks.

Like other perch-like fishes, cichlids have spiny rays in the anterior part of the dorsal, anal, and paired (pectorals in front, ventrals/pelvics below) fins. These hard spines help discourage predation, as the sharp spines make it difficult for a predator to engulf a cichlid, and they also serve as anchors or braces for the rest of the fin. Thus, the soft, segmented posterior rays of the dorsal and anal fins

Hemichromis fasciatus **is a substrate spawner. In general, the pairs of substrate spawning fishes tend to remain bonded for extended periods of time.**

can be used in various undulations for precise positioning of the body and for effortless movement through the water. In general, cichlids specialize in precision rather than in fast swimming.

Cichlids possess but one nostril on each side of the snout. Most other fishes have two sets of nostrils, with a set on each side of the snout. Water flows into the forward (anterior) nostrils and then out the posterior (toward the tail) ones after being sampled for olfactory cues. Since more advanced fishes don't have lungs, the nostrils are not connected to the respiratory system. Cichlids have simplified things by developing the single set of nostrils and at the same time improving upon them. The nostrils work something like syringes, the water being pulled in with muscular effort, sampled, and expelled. The advantage is that water can be sampled when the fish is not moving or if there is no current. Also, if need be, the water can be held in the nostril for a "longer sniff." This adaptation may enable cichlids to compete better in still water, and it is a feature that is shared by the damselfishes of the family Pomacentridae, all of which are marine but are generally thought to be closely related to cichlids.

Although not all cichlids have it and they aren't the only fish species to display

Geophagus steindachneri. **When the family feels threatened, the fry return to the safety of the mother's mouth.**

Both Discus parents will feed their young until they are able to fend for themselves. The parents secrete a specialized body slime that is optimal for the development of the fry.

it, a further advantage may be the broken and overlapping lateral line. The lateral line is possessed by most fish species and consists of pored scales with special nerve endings that enable them to sense water pressure. It is difficult for us to relate to this sense, but it is an important sensory apparatus for fishes that allows them to detect other moving things in the water and gives them feedback to their own movements. When scientists have disabled the lateral line by covering it in some way, fish move with stiffness and uncertainty about their environment. It is

believed that the broken lateral line may be an improvement that gives better feedback and covers otherwise "blind" positions.

The single set of nostrils, the broken lateral line, and the spiny rays all help nail down the identification of a fish as a cichlid. The only problem with all of this is that we would have to dissect a fish to make sure of the distinctions between damselfishes and cichlids. To determine the identity without dissection, the freshwater fish is likely to be a cichlid and the marine fish a damsel. It is true that

These juvenile *Neolamprologus leleupi* cave spawners from Lake Tanganyika have a local and plentiful supply of a small copepod that enables them to be self sufficient from the early free-swimming stage.

some damsels enter the mouths of rivers for short periods and some cichlids enter coastal salt water, but those are fairly rare exceptions. In any case, those features help us distinguish members of the cichlid family from all other freshwater species.

PARENTAL CARE

Cichlid behavior (especially the parental care of the young) must be mentioned as one of the secrets to the success of the family. Every single species known from the family Cichlidae provides some sort of parental care. This care ranges from substrate spawners that lay their eggs in a nest on the bottom and then guard the fry through a parent holding the eggs or young in the mouth during development (mouthbrooding) or even a combination of mouthbrooding and substrate spawning. It is oversimplifying things to divide cichlids into substrate spawners and mouthbrooders, but it is useful to do so.

SUBSTRATE SPAWNERS

In general, substrate spawners are those cichlids that pair up, clean off a spawning site, and lay adhesive eggs, usually on a bit of rockwork or driftwood. When the fry hatch out, the parents move them to a spawning pit that they have already dug. Normally the fry are helpless and unable to swim for about three to four days. During this time, the young are tended and guarded by the parents, and they move them from one pit to another. At first it was thought that this was done to keep potential predators guessing, but a more likely explanation is that moving the young means that they get cleaned of any microbes that might be injurious. During all this time the parents fan the eggs and the helpless fry to keep them free of debris and provide them with fresh water. The water around them should contain oxygen and not be over-burdened with carbon dioxide from the respiration of the young. Once the egg yokes of the fry have been absorbed and the young are free-swimming, the parents herd the young, all the while keeping them under their protection. During this time they stir up likely areas of the bottom that may contain food for the young. Many species secrete a specialized body slime that supplements the nutrition of the young. Both the male and female secrete this "milk." Discus are examples of cichlids in which the process is most extreme, the food being not supplementary but obligatory.

MOUTHBROODERS

With mouthbrooders, the male digs a spawning pit and the female is attracted to the site, laying her eggs in the pit, where they are fertilized. The female picks up the eggs and swims off to a safe area. The eggs are incubated in her mouth until the fry are-free swimming. The young then are released, but the mother still guards them for a number of days or weeks. In addition to chasing off potential predators, she takes the fry into her mouth for protection. There is not always a clear demarcation between mouthbrooding and substrate-spawning. For example, some geophagines (*Geophagus* and relatives) of South America brood the young in the mouths of both parents but release them when they are free-swimming and then tend them much like substrate-spawning parents. Mouthbrooders tend not to form pairs, and the female has sole custody of the eggs and young. In some species the male broods the eggs and the female guards him; other variants exist.

Whatever the mode, parental care has obviously favored cichlids as they have invaded new areas and radiated into many different species. There are enough differences between cichlids that scientists and hobbyists tend to specialize in certain groups, so it is worth our time to take a look at the cichlids of the world and just what they are like.

Kinds of Cichlids

The family Cichlidae has a relatively large distribution for a family of freshwater fishes, being found in the New World from southern Texas to southern South America, plus on the larger islands of the Caribbean. In the Old World they are abundant over the African continent, with a few species in

The Jewel Cichlid from central Africa spawns in the open. Both parents guard the brood.

The peaceful Uaru comes from South America. It is considered delicate, but Uaru are certainly worth any extra effort required in their care.

the Middle East and a handful in India and Sri Lanka; some distinctive forms occur on the island of Madagascar but are not closely related to those of Africa. Distinct types of cichlids occur in different places. The most profoundly evolved cichlids are in South America; discuses and angelfishes are examples of South American cichlid species that have evolved into greatly specialized forms. Cichlids don't dominate the fish fauna of South America, but they have certainly carved out niches for themselves. They do dominate in Central America, not in biomass but in species. It is believed that cichlids were one of the first families to colonize the area as it rose from the sea.

PRIMITIVE CICHLIDS

More primitive cichlids are found at the tip of India and on the adjacent island of Sri Lanka, as well as on the island of Madagascar. These cichlids may resemble the marine ancestor from which cichlids are believed to have descended. The area in which there are the most cichlid species is Africa; the main reason for this is that cichlids have come to dominate the Great Lakes of Africa, speciating wildly. Whenever you hear someone speak of African cichlids, they are nearly always talking about cichlids from one of the Great Lakes. The species from Lake Malawi are particularly colorful and active, so they have been especially popular with those aquarists who want motion and color in a large tank for the living room, den, or office. However, the cichlids of Lake Tanganyika have become very popular with advanced cichlid specialists and other hobbyists; the cichlids in that lake show much more variation in size and form because the lake is older and the cichlids have had more time to evolve to fill various niches. In Lake Malawi all but one of the known species care for the eggs and young via mouthbrooding. There are many different types of mouthbrooding in the lake, but there is none of the biparental care of the free-swimming fry for which cichlids are so well known. In Lake Tanganyika, not only is there biparental guarding of fry, but there is such extreme specialization among certain species that they are hardly recognizable as cichlids. By contrast, the cichlids of Lake Victoria are mostly derived from the single genus *Haplochromis* and have little variation in body type—even though many species are quite beautiful.

Generally speaking, the cichlids from Central America and South America show the most complex and intense brood care, but this is not a hard and fast rule, as *Hemichromis* and *Tilapia* species of Africa have some of the same intensity, while many cichlid species from Lake Tanganyika also demonstrate intense care. In general, the most colorful species are from Lake Malawi in Africa, but there are challengers not only

from the other African lakes but from Central and South America.

CHOOSING A CICHLID

A factor to take into account when choosing a cichlid is the type of water that comes from your tap. It is much easier to match species of fish to your water than to match the water to your fish. If you have soft, acid water, you may prefer to keep the cichlids of South America and West Africa, while if you have hard water with a high pH, the Central American cichlids and the cichlids of the Great Lakes of Africa are going to thrive in your tanks without altering the water. For those who want to keep cichlids of the Great Lakes of Africa (also known as the Rift Lakes) but have soft water coming from the tap, rift lake salts are available at your aquarium shop that will help you reproduce the water chemistry of the lakes and help maintain the pH of your tank.

Before you decide which cichlid you want to keep, it would be a good idea to review carefully some of the species that come from each area. The fact is that you can keep cichlids from a variety of areas as long as you take into account the levels of aggression and the types of water in which they prosper. It is perfectly all right to mix and match a few species from various areas. Such a tank can be quite attractive, since you have picked the best-looking fish species from each locale. Even if you go that route, it is a good idea to know the origin

This is a male Frontosa from Lake Tanganyika. It is a mouthbrooding species. The female, when holding the brood, will eat for herself and the babies.

West African cichlids, like this *Pelvicachromis* species, are often very colorful.

of different species so you can explain it to company.

While the cichlids of Central America are quite variable, they couldn't hope to compete with the ones from South America; those have been there for a long time and have evolved to perfectly fit many niches, from the predatory *Cichla* to the omnivorous *Uaru*. The very popular discuses and the angelfishes come from this region, as do many types of very specialized cichlids. In general, most South American cichlids prefer soft water, although there are exceptions to that rule. Their spawning behaviors range from substrate spawning to mouthbrooding to a combination of the two. Their feeding systems and life histories are also quite diverse.

The most primitive cichlids are found on India (and Sri Lanka) and on the island of Madagascar. These cichlids are primitive as it is believed that they most closely represent the ancestral cichlids and have remained that way because they have not had to compete with modern cichlids.

One of the more neglected groups today is the cichlids of western Africa, as more keepers look for the species from the Great Lakes in eastern Africa. West African cichlids often are extremely colorful, as with the various *Pelvicachromis* species, but some are more noted for their aggressive behavior (*Hemichromis* species). Many are dwarf cichlids and easily kept in small aquaria, where they will spawn happily for years.

Even in the smallest town, the average hobbyist probably has access to several species of cichlids from around the world, including a variety of sizes, colors, and prices. With a little care, even the beginner can choose an interesting, attractive cichlid species that will live for years in the aquarium and may even spawn.

The Cichlid Tank

These filters are extremely good at polishing the water. They do so by forcing it through a layer of the silica remains of long-dead minute unicellular algae.

FILTRATION

A lot of thought should be given to filtration. For many reasons, the undergravel filter is one of the most popular of all filters, but it can be a problem with cichlids because most cichlids tend to dig holes in the gravel. This is particularly true when spawning time arrives. Holes completely subvert the action of the undergravel filter. What happens is that the cichlids dig all the way down to the filter plate, exposing it. Since water tends to flow through the course of least resistance, all the fluid tends to flow through the area of the exposed filter plate, completely bypassing the gravel, which was intended as the filter medium. Hence, most cichlid keepers utilize some type of filtration other than an undergravel filter. It is possible to use an undergravel filter with cichlids, but some modifications have to be made. A common solution is to put in the gravel, put a piece of grating

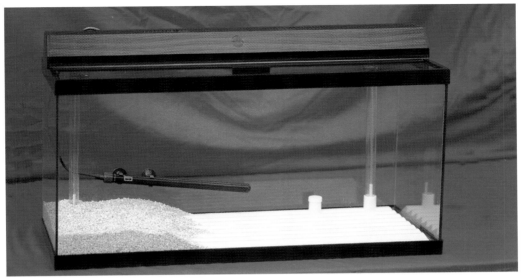

The undergravel filter, a favorite for many years, is not generally recommended for a cichlid tank.

(such as that used as fluorescent light diffusers) on top of the gravel, then add more gravel on top of that. When the fish try to dig, they can't get down to the filter plate through the grating. Other modifications include using plastic screening, which simply keeps the fish from digging below that level. The fact is that these alterations are not used by keepers of large cichlids, as those fish could very likely pull loose the grating or screening.

Even modified, the undergravel filter is generally an unsatisfactory arrangement for cichlids. Most cichlids are of a larger size than typical aquarium fishes and thus dirtier, so the gravel needs vacuuming more frequently. It is difficult to get the gravel clean with grating or screening in the way.

MECHANICAL AND CHEMICAL FILTRATION

One reason that the undergravel filter has

Carbon removes many impurities from the water and is particularly useful when the fish produce a lot of waste, as larger cichlids do.

Good filtration buys time. Regular water changes are important. This homemade trickle filter complements the work of the aquarist in maintaining optimum water conditions.

been popular is that it takes care of two aspects of filtration, mechanical and biological. Mechanical filtration refers to the clearing of the water of debris. Chemical filtration refers to removing dissolved compounds by means of absorption (into the body of a filter medium) and adsorption (on the surface of the medium). Biological filtration refers to the breaking down of ammonia into less harmful compounds. Until recently, that consisted of converting ammonia (resulting from decomposition of organic matter from the fish and uneaten food) into nitrite and breaking down nitrite into nitrate. This is all done by bacteria of different types. Ammonia and nitrite compounds are quite harmful to fishes, but nitrate can be tolerated in low levels. Unfortunately, the only bacteria that

break up nitrate are anaerobic, functioning in the absence of oxygen, and part of their respiratory system involves giving off toxic substances. Constructing a filter that breaks down nitrate has been difficult and fraught with hazards to the fish.

Mechanical filtration usually involves the use of paper, plastic, or sand (or all three) as the medium. Also used is diatomaceous earth, which consists of the silica "skeletons" of tiny diatoms. Chemical filtration usually is done with activated carbon but also may involve ion exchange resins. Generally speaking, a filter gets added points if it has all three types of filtration. This is because the object of filtering is to remove debris from the water and also get rid of harmful substances that are dissolved in the water.

26

WHICH FILTER IS BEST FOR CICHLIDS?

With the foregoing in mind, there are several filters that can be recommended for cichlid tanks. One of the old standbys is also the simplest: the inside box filter. Powered by air, it runs the water through the medium quickly and utilizes all three types of filtration. The filter normally contains activated carbon for chemical filtration and some sort of synthetic floss for mechanical filtration. Eventually the carbon and floss are colonized by desirable bacteria so that biological filtration is also present, albeit weakly. A good strategy with box filters is to have several of them in a tank and clean a different set every two weeks. That way the bacteria will remain operative in the filters not yet cleaned and will have a chance to build up in the newly cleaned filters.

Part of the decision of what filter to use is determined by what the hobbyist wants the tank to look like. In breeding tanks, inside filters (which tend to make an aquarium appear cluttered) are perfectly acceptable because no one is going to make a breeding tank look like a show tank. Even in a show tank, cichlids are going to breed. (Just try to stop them!) However, in a show tank we are more concerned with appearance, so we may not like the looks of a number of inside filters in the tank and may prefer to use a canister filter. These are quite efficient and provide filtration under pressure. Please be advised, however, that cichlids are active fishes of fairly large size, and they are going to provide lots of metabolic wastes. For that reason, use a filter that

A powerhead with a prefilter can quickly clear a lot of floating detritus from the water. If the prefilter is cleaned often, i.e., not thought of as a biological filter, it can remove waste before it dissolves.

Fast-swimming agile fish, like this giant danio, are ideal for inclusion in some cichlid tanks. They can be excellent dither fish.

can be changed at least every two weeks. If you get a canister filter, get one with a lift-out basket for easy changing or cleaning of the medium. Activated carbon often is a component of the filter media in a canister filter, so such canister filters provide all three types of filtration and do so in an efficient manner.

Outside power filters fit on the side of a tank. They are easily serviced and can have enough different filter media to provide all three different types of filtration. Those with a rotating section (bio-wheel) are particularly adept at providing biological filtration. A bio-wheel provides wet-dry filtration in a compact way. Wet-dry filtration is an attempt to provide the bacteria with the oxygen that they need to do their good work. The bio-wheel provides that by a large rotating wheel which dips into the water and then rotates into the atmosphere, thus exposing a large sur-

face of the water to gas exchange from the atmosphere.

Trickle or wet-dry filtration created a sensation when it was introduced. It consisted of a tower in which some sort of suitable habitat for bacteria was housed. Usually this consisted of "bio-balls," small plastic balls with multitudes of ridges and crevices in order to provide lots of habitat for the bacteria. The water is pumped out of the tank through a prefilter (for mechanical filtration) and then allowed to trickle down through the bio-balls for the dry part of the filtration. The water then collects in a sump, from which a pump propels the water back to the tank.

Although one of the ultimate biological filters, the wet-dry filter is expensive, and it has not been widely used in cichlid aquaria because cichlid people tend to emphasize regular partial water changes. The fact is that it was cichlid people who pioneered the idea of regular partial water changes. The old idea was that there was something magical about old water. Cichlid and marine hobbyists helped break that tradition when the tanks in which regular partial water changes were made had better success than those that had no regular partial water changes. This was true regardless of filtration method.

The lesson may be that filtration isn't that important in the cichlid tank, but I think it is a mistake to downplay filtration. Good filtration buys us time: our water stays in good shape for a longer period of time. Most of us make trips and don't want to assign water changes to whoever looks after our fish. We also never know when something is going to so command our attention that we are less than exemplary about keeping to a water changing schedule. One bit of apparatus that will make water changing easier, and thus more likely to be done, is the automatic water changer that attaches to a faucet. Not only does it fill the tank with water, but it also removes it. It is easily attached and deployed and sure beats having to carry five-gallon buckets of water!

OTHER FACTORS

The cichlid aquarium will be different from other aquaria in several respects. First, depending on the cichlid species being kept, tankmates must be selected with care. Non-cichlid fishes should be of the type that inhabit the upper part of the water and are fast-swimming and

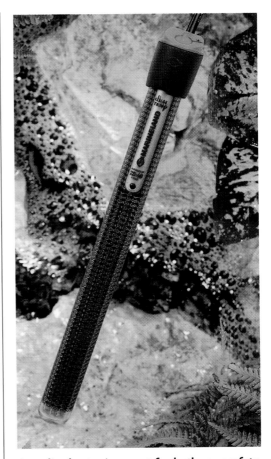

A quality heater is a must for both your safety and the well being of the fish. Some filters have an enclosure for the heater. This keeps it out of the way of marauding cichlids.

agile. I know that people become so taken with cichlids that they put together community tanks composed strictly of cichlids, but I want to emphasize that any fishes that are kept with cichlids should be selected with care. Plants, unfortunately, are going to be out, but you can compensate with colorful rockwork and driftwood as well as plastic plants. Yes, I know that some hobbyists have succeeded in keeping some plants with cichlids, but these were people who really understood their fish species and the plants. I don't

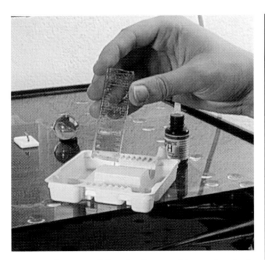

In a perfect world nobody would need a test kit. In the here and now the use of a kit is your only early warning system. Without a kit, your first hint of trouble may well be a cichlid corpse.

want to put anyone in the position of being unhappy with their cichlids because the fish uprooted and tore asunder a valuable Amazon sword plant. Most cichlid species are not even plant eaters, but the plants take a lot of punishment, most particularly when cichlids are in the process of spawning. Nature has endowed most cichlid species with an abundance of aggression for the purpose of the protection of the eggs and young. The fish tear around the tank looking for potential enemies of their young, and the plants get uprooted and ripped as the result of cichlid anger.

Naturally, you need to use special compounds for breaking down chlorine and chloramine in the water, as most water is treated to make it safe for human consumption. Sensitivity to chlorine varies greatly with species and individual cichlids, but just storing the water in a con-

tainer with a large surface area should allow most chlorine to disperse. If attempting to modify your water chemistry (hardness, pH, etc.), try to use as few chemicals as possible and in the smallest possible quantities. All water should be the cleanest possible when it is put into the aquarium.

Aside from ornaments, filtration, and tankmates, a special consideration for cichlids is the type of gravel employed in the tank. Most cichlids like to dig in the gravel, so there is little worry about anaerobic spots developing in the gravel without an undergravel filter in the tank. My recommendation is to utilize a fine grade of gravel so it is easier for the cichlids to dig, but only provide them with a relatively shallow layer of about 2 to 3 inches. You will want to be able to vacuum out the gravel bed with some regularity, so you don't want gravel that is too fine.

Returning to the subject of tankmates for just a minute, it is worth mentioning that the same type of people who like cichlids also tend to like catfishes—especially of the family Loricariidae (suckermouths). The bad part about this situation is that cichlids and catfishes both tend to like to stake out claims on the bottom of the aquarium. You don't need catfish for cleaning up in a cichlid aquarium, as most cichlids are quite adept at picking food from the bottom, and many species keep the gravel cleaned off pretty well, too. My recommendation is to leave catfishes out of a cichlid tank.

In general, most modern aquarium tanks are going to be good cichlid tanks. Modern tanks may be made with regular glass (breakable, heavy) or acrylic plastic (expensive, scratchable); the choice is one of personal preference. The lighting of a cichlid tank is also one of taste, but it is worth bearing in mind that, though some cichlids look good in bright light, most are at their best in slightly subdued lighting. With that in mind, you may want to consider a light with a rheostat so that you can vary the intensity of the lighting. Since you won't be keeping plants, there is no reason to have the lights on when no one is around to enjoy viewing the tank; protracted periods of light will just encourage the growth of algae.

Most cichlids come from tropical areas and thus require warm water to thrive. In most setups a heater will be necessary. Both submersed and hanging heaters work well and are reliable, though larger cichlids may learn to hit heaters (perhaps detecting the electric current within) and will need exceptionally sturdy tubes around the heater. A temperature between 78 and 84 degrees Fahrenheit works for most cichlids. Of course you also need an easily readable, reliable heater.

FOOD

We are fortunate that these days there are many good prepared foods for cichlids. Any good aquarium shop should have an interesting selection of floating and sinking foods, large and small granules, flakes, colorfoods, high or low plant matter content, and price ranges. An occasional feeding of live brine shrimp or bloodworms is great, but it is not a necessity, as nearly all species of cichlid prosper on commercial foods, both dry and frozen. The only choice you need to make involves the size of the flakes or pellets that you feed, and that choice will be dictated by the size of your species.

Once you have set up the tank and given the filter enough time to begin functioning, it is time to look for your fish. Of course you already decided in a general way what type of cichlid (size, water chemistry) you will be keeping before setting up the aquarium, but now you get the chance to make you selection of species by color and availability.

Cichlids of the Americas

CENTRAL AMERICAN CICHLIDS

Just so you will be familiar with the term, ichthyologists often refer to animals of the New World tropics (the Americas) as "neotropicals." All the fish we discuss here would technically be referred to as neotropical cichlids. Also, older, more recognizable generic names have been

Convicts make devoted parents. They like to lay their eggs on the walls or roofs of caves and produce between one and two hundred eggs at a time.

The Copper Cichlid is available in many color morphs, all of which breed freely. This results in a good aquarium fish that presents a spectrum of colors and patterns.

used for the major listing, with newer names (splits) placed in parentheses and used as though they were subgenera. Scientific names of the cichlids are subject to rapid change, with many specialists using different generic names for the same fish. Expect to eventually see most of the parenthetical names used as genera in some books and articles.

SMALL SPECIES

Small is a relative term of course, but I thought that I would present the species broken down by size. That doesn't mean that only small cichlids should be kept together. Some of the small ones are so aggressive that they dominate some of the larger cichlids. I will try to take special notice of the species that can be mixed with big cichlids. I should mention that none of this is absolute, and some strange combinations have been made under unusual circumstances.

There are well over a hundred Central American cichlid species, so obviously I have been able to include only a few of the most common species. Note that currently the genus *Cichlasoma* is technically restricted to a few mostly South American species, so quotes are used around the generic name for the Central American species.

Convict Cichlid

"Cichlasoma" (Archocentrus) nigrofasciatum: This is a cichlid that has often been called the "missionary cichlid," as it so often entices a neophyte into the cichlid hobby. It is not that the fish is beautiful,

but it is unerringly a good parent—and on the very first try, too! Further, it spawns at a young age and seems to be in the process of either spawning or raising young during most of its life. A further favorable attribute of these cichlids is that they spawn at a very small size, with the male barely over an inch long and the female barely reaching an inch in length.

Though quite common, they never lose their appeal. A recent innovation among cichlidophiles is to try to collect the several color variations. Since this species is spread over a wide area of Central America, there is considerable variation in coloration depending upon the locality where the specimens were found. For instance, there are especially red color forms that are found in Honduras. In these specimens, the red is found on both males and females, but on the female it covers more of the body, producing a flaming belly region and red all the way up to the dorsal fin. In the more common types the male shows no red, but the female sports copper coloration in the ventral (belly) region.

The Convict feeds on the bottom, picking up foods opportunistically. The food can vary from invertebrates inhabiting the substrate to a mixture of organic (plant and animal) matter. The parents feed the young off a specialized body slime that is secreted by both parents. This behavior of supplying a supplementary "milk" is very common among Central American cichlids. The parents generally supply food for the young by chewing up food that is too large for them and spitting it out in the school of young as well as "sweeping" the bottom with their anal fins to stir up any possible food. All these behaviors, though typical of many cichlids, are what have endeared this particular species to aquarists around the world.

Copper Cichlid

"Cichlasoma" (Archocentrus) septemfasciatum: This is a cichlid that changes dramatically when spawning. The popular name comes from the coppery coloring that often covers the body of the female. Something that all of the *Archocentrus* group share is that the female is pretty much in charge of the eggs before they hatch, while the male guards the perimeter. Once the eggs hatch, the male aids in moving the young to a pit; several holes are dug during the four days before the young become free-swimming.

When the species spawns, the female develops a dark mask that looks very much like a knight with his visor pulled down, ready for battle. While not a knight, a guarding Copper female is truly ready for battle!

Golden Cichlid

"Cichlasoma" (Thorichthys) aureum: There are several popular names for color variations of this quite charming cichlid, including "gold flash" for one of the variations. The most popular aquarium species of this group is the Firemouth Cichlid (*Cichlasoma meeki*), but all of the Thorichthys group greatly resemble one another. They are substrate sifters, but they don't run the gravel out through the

Often called the Gold Flash and with good reason, this fish is speedy and spangled with iridescent scales.

gills, instead turning it over in the mouth and then spitting it out. In the wild they feed primarily off invertebrates that they find in the bottom, mostly insect larvae. In the aquarium they take food opportunistically, and they are good feeders and hardy aquarium fish.

The species of the *Thorichthys* group really push the "small" cichlid category, as in several species the male reaches a length of 6 inches, with the female about 5 inches, but these cichlids are quite mellow as compared to cichlids in general. True enough, they are fierce protectors of their young, but they are not as capable of inflicting damage on other fishes as are many cichlids, and when not spawning they are more quiescent than many cichlid species. I have kept them with angels, for example, something that could not be done with most other cichlids.

Firemouth Cichlid

"Cichlasoma" (Thorichthys) meeki: This species has long been a favorite in the aquarium, even though it is not suitable for the typical community aquarium. The reasons for its popularity are obvious. It

has a distinctive shape, and its coloration is quite pleasing. It has always been one of my favorites, and the fact that it is common has never bothered me. Apparently it hasn't bothered other hobbyists either, as it has remained consistently popular. The fact is that it can be kept in a community aquarium until spawning time arises, when the other fishes will be driven off mercilessly. The fish spawn when the male is a mere 3 inches, with the female being about 2.5 inches in length. However, the male can eventually reach nearly 6 inches in length. As compared to many other cichlids, the pairs seem to stay together well in the home aquarium, with no spats between spawning.

This species' bluff may be even more impressive than that of the others in this group. During the frontal threaten display the red coloration is more intense in this species, especially toward the head. For some reason, red coloration is also intimidating to fishes, a fact discovered by ethologists.

There are color differences in this species depending upon the locality. The species ranges from southern Mexico to Guatemala, being found primarily on the Yucatan Peninsula. Some specimens have more blue spangles, others more emphasis on the red. The most beautiful, and therefore the most popular, specimens are those that show a balance of bright red with lots of blue spangles.

Rainbow Cichlid

Herotilapia multispinosa: This fish is a candidate for the smallest of the Central

American cichlids, with the males reaching 3 inches or only a little more. This is a species that seems capable of defending its young without committing total mayhem on its tankmates. This fish is distinctive from other Central American cichlids in that it has tricuspid (three-pointed) teeth that enable it to more easily harvest filamentous algae, which make up a good portion of its diet. In the aquarium it will take all foods, but some vegetable matter should be a part of its diet, even if it is only dry foods designed for herbivores.

Poor Man's Tropheus

Neetroplus nematopus: This species is somewhat analogous to the Tropheus of Lake Tanganyika in that it lives off the algae that grow on the rocks in lakes and rivers. It also somewhat resembles Tropheus. They are one of the easiest cichlids to spawn, and they are almost impossible to stop once started. They run the Convict Cichlid a close second in propensity to spawn and efficiency at protecting their young. An interesting

Neetroplus nematopus, like the Tropheus species of Lake Tanganyika, lives off the algae that grow on the rocks where it is found. This fish reverses colors at spawning time.

aspect of this fish is reversal of colors at spawning time: the dark stripe on the side becomes quite white and the normally pale body becomes nearly jet black. This fish is found from Costa Rica to Nicaragua in lakes and rivers, including the Great Lakes of Nicaragua.

MEDIUM SPECIES

Some of these cichlids may be quite large, most of them eventually reaching over 6 inches, especially the males, but many are quite gentle. Others have evolved a quite truculent temperament in order to protect their young. Pairs of such species may be difficult to keep together in a tank unless the tank is of a sufficiently large size. If the tank is not big enough, horrific fights may erupt. Since the female is considerably smaller than the male, she usually gets the worst of it and may be killed. Prudent cichlid hobbyists with insufficiently large tanks use a glass partition between the two and have it propped up about a quarter of an inch to allow the male to fertilize the eggs. The fry can then be left in with the parents, as the cichlids—versatile critters that they are—manage to act as though the entire thing is a normal spawning event.

Red Devil

"Cichlasoma" (Amphilophus) labiatum: When these cichlids first arrived in the country, dealers nearly put them in ocean water, as they resembled the Garibaldi, a large California damselfish, in color. Exporters had actually mixed up Red Devils with red Midas Cichlids (*"Cichlasoma"*

The Red Devil is an aggressive, capable fish that should be kept by itself in a large aquarium. In the wild many individuals remain the gray color they were as juveniles.

[Amphilophus] citrinellum). For that reason, the Midas Cichlid is still called a Red Devil in some circles, but it tends to be golden, rather than red, and it has a less pointed snout. The large lips of the true Red Devil tend to regress under captive conditions for some reason that is not completely understood. These cichlids were not named Red Devils without reason, as they are quite aggressive and have the dentition to back it up. Nothing could seemingly be kept with them, but it was later discovered that they could be kept (in very large tanks, of course) with a number of other cichlids while growing up.

With all of these *Amphilophus* cichlids, the best way to breed them is to get about six of the juveniles and allow them to pair up naturally. Of course, once a pair spawns you will have to remove the others—unless you have a 2000-gallon tank! The juveniles are gray in coloration and then begin to turn speckled before they turn red. Interestingly enough, the red coloration is quite variable, and the fins are often fringed with black (the most beautiful pattern, in my opinion); in the wild many individuals stay gray.

The male can reach nearly a foot in length, but it takes about three years to attain such a size. Pairs will spawn when only 6 inches long.

Jack Dempsey

"Cichlasoma" (Amphilophus) octofasciatum: You can tell by the popular name of this fish how long it has been in the hobby, as Jack Dempsey was the heavyweight boxing champion in the 1920s. Most people these days simply refer to this long-time favorite as "Dempseys." The fact is that the fish is not nearly as aggressive as the Red Devil, for example, but it is durable and tough enough to be kept with such fish in a tank that is reasonably large. The species is found on the Atlantic slope from Mexico to Belize.

Juveniles are mostly black, without the blue spangling that really "makes" the appearance of this fish. It takes patience to see this fish at its best, as it takes slightly over a year to fully mature, males reaching 8 inches. Specimens barely 3 inches long may breed in captivity.

Like other Central American cichlids, it prefers slightly hard and alkaline water, but it is quite adaptable.

The Jack Dempsey is well known for its ability to defend itself, but most people don't realize what a stunning fish it is in its adult manifestation.

The Parrot Cichlid is a beautiful Central American cichlid. The females practice an unusual system of brood care. Termed creching, a group of three or four of them stands guard over their combined spawns, encircling them and repelling any would-be intruders.

Parrot Cichlid

"Cichlasoma" (Copora) nicaraguense: Although males can reach a length of 10 inches, this is a rather gentle cichlid as compared to the others, and it certainly is in the running for the most beautiful Central American cichlid. Females are actually more colorful than males, but the males are certainly not somber in coloration; they simply have less gold and more emerald colors than females.

The species is found in the Great Lakes of Nicaragua as well as in numerous lakes and rivers into Costa Rica. Its behavior is distinct in many ways. For one thing, it inhabits sandy areas and the fish spawn in a sandy pit, producing bright golden eggs. These eggs are non-adhesive and actually bounce around to some extent when the female fans them. Additionally, the female will form a nursery territory into which no predators are permitted. Such behavior is not often seen—even among cichlids! This is one of those larger species that can easily be kept with smaller cichlids, but it should not be kept with some of the tougher ones, such as Red Devils and Texas Cichlids.

Texas Cichlid

"Cichlasoma" (Herichthys) cyanoguttatum: In the aquarium hobby this species often has been known as the Pearl Cichlid due to the bright pearly spangling on the front part of the body when in spawning color. A species of northeastern Mexico, it also ranges naturally into southern Texas. The Texas Cichlid is durable and tough enough to be kept with the Red Devil. Males commonly reach a length of

The Texas Cichlid, also known as the Pearl Cichlid, lives farther north than any other member of the cichlid family.

Herichthys carpintis is even brighter in color than the Texas Cichlid. It comes from Northeast Mexico. Many color variants exist.

7 to 8 inches and may develop a large hump on the head. The very similar Blue (or Green) Texas Cichlid, *"Cichlasoma" carpinte*, is even brighter in color (with more pearly spangling on the face) and also often seen; it comes from northeastern Mexico.

Tricolor Cichlid

"Cichlasoma" (Nandopsis) salvini: This is a beautiful but very aggressive cichlid from the Atlantic slope of Mexico to Guatemala, where it is quite common in Lake Peten. The female is more colorful than the male, but both of them can look quite lovely. These fish are predators upon large invertebrates and small fish, including the fry of other cichlid species. They mix well with Dempseys, and they can also be kept with Red Devils. Of course, all bets are off when spawning time arrives, as whichever species has the fry is likely to dominate. Males reach a length of about 8 inches, but the fish spawn at a much smaller size.

These Tricolor Cichlids come from fast-flowing rivers in Mexico. The males have bright green spangling and the females are even more beautiful in their bright yellow and red garb.

Orange Tiger Cichlid

"Cichlasoma" (Nandopsis) urophthalmus: Found on the Atlantic slope of Central America, this very adaptable fish also inhabits mangrove areas with salt water. The fish work the roots of the mangroves in search of small fish and large invertebrates. They are good feeders, but they are among the most aggressive cichlids, so plan accordingly and keep them with fish with similar levels of aggression (Red Devils again!).Males can easily reach 9 inches in length.

ROUGH AND TOUGH

These guys are pretty rough, and since they get so large, the only way to spawn them often is to utilize a divider in the tank, either a glass panel raised a quarter of an inch from the bottom or a grating made from a plastic fluorescent lighting diffuser. People who keep these fish claim that they are worth the trouble because of their interesting behavior and sometimes beautiful coloration.

A common way to keep these fish is to keep just one as a kind of pet. They are such intelligent fish that they react with their keepers, much like a dog or a cat. Hobbyists even provide them with toys, such as floating ping-pong balls, and this is not a bad idea. Any large cichlid kept by itself should have at least rocks and tunnels with which it can indulge itself.

Wolf Cichlid

"Cichlasoma" (Nandopsis) dovii: This predator has such powerful jaws that if a fish is too big to swallow, it consumes it

The Wolf Cichlid is a savage predator. It is very intelligent and the largest of all the cichlids.

by ripping it to pieces. Rarely is it possible to keep one with another fish, even a large armored catfish, as it simply won't tolerate another fish within the confines of a typical aquarium. Wolf Cichlids are about as rough and tough as they come, yet they are extremely intelligent. Breeding in captivity involves separating the parents with a partition; in the wild the parents cooperate wonderfully in order to protect the young. Males reach nearly 2 feet in length.

Jaguar Cichlid

"Cichlasoma" (Nandopsis) managuense: This species is found on the Atlantic slope from Costa Rica to Honduras and is about

The Jaguar Cichlid preys on small fish and large invertebrates. It is the second largest species, after the Wolf Cichlid, in the genus.

half the length of the Wolf Cichlid. It can be quite fiery, friendly with their owners but totally pugnacious with strangers. There is a lot of individual variation among the different specimens. Some individuals will even be slightly shy, but this is rare. The bluish silver color heavily spotted with dark brown is quite distinctive and attractive.

Trimac

"Cichlasoma" (Nandopsis) trimaculatum: This red-eyed cichlid practically breathes fire. When James Langhammer was curator of the Detroit Zoo, he wrote of Trimacs that took feeder mice away from the alligators! This is another cichlid that delights those people who like rough fish. They are from the Pacific slope of Mexico. Trimacs are good parents. The best way to get a pair is to obtain six to eight young, raise them together, and let them pair off naturally. The male reaches a length of about 14 inches.

Basketmouth Cichlid

"Cichlasoma" (Nandopsis) umbriferum: This is an American cichlid whose range extends from Panama to Colombia on the Atlantic slope. A beautiful pale tan with rows of bright blue spots, the Basketmouth competes with the Wolf Cichlid for the title of largest species of Central American cichlid, some males reaching 16 inches. It inhabits open water and preys upon schools of livebearers and tetras that inhabit the area.

This species seems less aggressive than *Cichlasoma dovii*, but they are plenty aggressive, and they have the size and dentition to back it up.

Black Belt Cichlid

"Cichlasoma" (Theraps) maculicauda: This is one of the truly gentle giants of the cichlid world. The most widely distributed cichlids in Central America, inhabit-

Due to their ability to tolerate brackish and marine conditions, the Black Belt Cichlid has the widest distribution of all Central American cichlids.

ing coastal lagoons and estuaries as well as lakes and rivers, they are quite colorful fish. The adults can be kept together without any problems, especially if you raise about six young together and let them pair up naturally as they mature. The male reaches a length of a little over a foot, but these fish have very deep bodies.

SOUTH AMERICAN CICHLIDS

To make a place for themselves in one of the vast drainage systems of South America, cichlids had to find niches that were previously little exploited. It is believed that they had at least 50 million years in which to evolve, finding unexploited or little-used feeding niches in South America and then moving from them to areas of extreme specialization. Among the most modified are the round and extremely compressed discuses (*Symphysodon*), which live in areas with almost no food for baby fish to eat; both parents secrete a specialized mucus on their sides for the young to feed upon. So bizarre and different are discuses and the somewhat similar angelfishes that they are hardly recognized as cichlids by general fish hobbyists. Although discuses and angelfishes are extreme variations in cichlid body type, there are other specializations that have evolved in the very competitive environs of South America, including the bottom-sifting eartheaters or geophagines and the very predatory pike cichlids, to mention only a couple. A candidate for the largest cichlid in the world lives in South America, the Tucanare or Peacock Cichlid (*Cichla ocellaris*), a large predatory form often exceeding 2 feet in length.

Of course not all the cichlids of South America are oddly shaped or giants. Most are of familiar cichlid form and under a foot in length, with many species of dwarves that remain under 4 inches all their lives. The dwarf species are popular with many hobbyists, though they seldom are easy to find in shops; adult males commonly are spectacular little gems, with bright colors and long fins. Many of the odder South American cichlids are available only through specialist importers or breeders and have a strong following among advanced keepers. The following species offer merely a taste of the fascinating cichlid fauna of South America.

COMMUNITY TANK SPECIES

Spadetailed Apisto

Apistogramma agassizi: There are dozens of species of *Apistogramma*, but many of them are relatively colorless and only appreciated by the real apisto devotee. This staple of the tropical fish hobby occurs throughout the Amazon, right up to the base of the Andes Mountains, and long has been popular. Like other *Apistogramma* species, spawning females turn yellow, guard the eggs by themselves, and take primary care of the young, with the male patrolling the perimeter of the territory. Even during a spawning, the fish are unlikely to harm other inhabitants, as these are such mild fish that a female guppy may challenge the fry-guarding female. In nature, most of the *Apistogramma* species are harem-spawners, with one male presiding over the territory of several females, so it may be best to keep one male with several females in a community tank. Apistos are found in soft water in the wild, so it is advisable to provide soft water for spawning at least, but they can prosper in hard water if you aren't worried about breeding them. The young are able to take newly hatched brine shrimp once they are free-swimming. Males reach about 3 inches in length, with the female half this; they spawn at a much smaller size. Spade-tailed Apistos and others of the genus are short-lived for cichlids, living three to four years.

When guarding eggs or fry, the Blue Ram manages to be present at all the potential trouble spots around the boundary of its territory, almost it seems, at once. Somehow it can give an impression of perfect stillness at all of these spots. It is like a self-important hummingbird.

Ram

Microgeophagus ramirezi: This Venezuelan fish was named after the collector, Manuel Ramirez, but the scientific name was butchered to become the popular name "Ram." There is a golden variety, but the natural blue color is truly beautiful. Although the colors are somewhat subdued except when spawning, the fish has a real touch of class.

In spawning (soft water), both parents clean off a rock and both of them care for the young, just like some of the large cichlids. However, they are only about 2.5 inches long at maximum and are quite gentle, so they are no threat to the other members of a community tank. Rams have been favorite aquarium fish for almost half a century. *Papiliochromis* and *Mikrogeophagus* sometimes are seen as the generic name.

Golden-eye Dwarf Cichlid

Nannacara anomala: This species is very similar to *Apistogramma* species in behavior, but it is considerably different anatomically. They may appear transparent brown under bright light, but when in good condition and subdued lighting the male has a nice bluish iridescence and rows of dark spots can be seen on the sides; females have two brown stripes on each side. These fish spawn in an identical manner to apistos except that the female does not turn yellow. This is a cichlid that can spawn in a community tank without terrorizing and killing the other inhabitants, although they will be chased away from the spawning area. It comes from the Guianas in northeastern South America.

The Golden–Eye Dwarf Cichlid lays its eggs on a flat stone and after two days the young are helped out of the shells and moved to a series of pits in the sand.

Common Angelfish

Pterophyllum scalare: This exotic fish is thought of as common now, but it certainly created a sensation when it was first imported early last century. They were expensive (they had to be shipped in large containers by freighter all the way up from the Amazon) and were not easy to spawn, but their captive-bred descendants are not so difficult. Unlike most other cichlids, angelfishes like to stay up in the water col-

Tankmates for Angelfish should be carefully chosen. Their fins are magnets that draw many to harass them. If this happens the fish will never look its best.

umn (preferably in dense vegetation or branches), although they will pick up food from the bottom, looking unbelievably stately as they do so, the long, flowing fins and vertically barred pattern making them unique. (There are at least three almost identical species.)

Angelfishes spawn on plants or tree limbs in the wild. When the young hatch, they are moved to other plants (instead of putting them in pits) and later are transferred from plant to plant. When the young are free-swimming, they will take newly hatched brine shrimp but still are cared for by both parents. Unfortunately, Common Angelfish are raised commercially by removing the eggs from the parents, so many may not be effective parents. The sexes are practically identical in appearance, and even experts may be wrong almost half the time when sexing them. Many fancy varieties have been bred in the aquarium, including gold varieties, but the natural silver and black form still looks the best to many keepers.

Discus

Symphysodon aequifasciatus: Along with the Common Angelfish, this species is generally considered the most exotic of tropical fishes, but many cichlidophiles would disagree. There are believed to be at least two species of discus (the other generally accepted species is *Symphysodon discus*), and there are two types of discus hobbyist. The most common are the ones who breed fancy varieties, of which there are dozens named, though many (in my opinion) are not nearly as attractive as the wild type. Other aquarists want their discus as close to the wild types as possible, and they even make expeditions down to the Amazon to capture specimens for diversity in the captive genetic pool.

Discus prove difficult to keep for many aquarists, as they need large tanks and prefer a diet based on live and frozen insect larvae such as bloodworms and mosquito larvae. They must be kept in soft, acid water held at 80 to 85 degrees. The sexes are virtually impossible to distinguish externally, so most hobbyists buy a group of young specimens and let them pair off naturally as they mature.

CICHLIDS FOR SPECIAL TANKS

The following cichlids are not quite mild, but they aren't typically cichlid aggressive either. They can be kept in a cichlid-only aquarium or with larger fishes that can get out of the way if the cichlids decide to enforce a territory.

The Threadfinned Acara is a harem spawner with up to half a dozen females to a male. The females often spawn at the same time.

Threadfinned Acara

Acarichthys heckelii: This is a relatively hardy South American cichlid that can tolerate a wide variety of water conditions and feeds well on a variety of foods. Males have long red filaments on the posterior part of the dorsal fin; large adults may be 8 inches long. Spawning this species in the home aquarium is not simple. In the wild females dig chambers in the clay of the river bottom or banks, creating a maze that predators must enter to reach the fry. A female actively solicits a mate once the chambers are ready, but she only wants him for fertilizing the eggs, after which she drives him off. In the wild, the male presides over a harem and guards a large territory that includes all his females. Some hobbyists have had spawning success using lots of inverted clay pots of various sizes or using various sizes of PVC piping, maintaining a high temperature (85 to 90 degrees) along with lots of aeration and frequent partial water changes.

Blue Acara

"Aequidens" pulcher: The genus *Aequidens* has been restricted to larger species, and this species is excluded from the genus, but no new genus has been erected for it and the many species like it, so the quotes make known the fact that the genus for this species will probably eventually change.

In any case, the Blue Acara has been a popular fish in the aquarium hobby for over 60 years. One reason for its popularity is that it is only about 5 inches long. The fish are excellent parents, with both of them caring for the fry, are ready spawners, tolerate a variety of water conditions, and don't demand a huge territory, even when spawning, so they aren't overly tough on tankmates. The typical form comes from northwestern South America and Trinidad.

The Blue Acara continues to be a popular fish that breeds readily in captivity. Males grow larger than females and turn blue in the mating period.

The Cupid Cichlid is a peaceful fish that becomes less so only at spawning time. It demands good water quality.

Cupid Cichlid

Biotodoma cupido: This species was described as a species of *Geophagus* but has technical characters distinguishing it from that genus. Cupids can reach about 6 inches in length, but they are quite peaceful, causing problems mostly when spawning time comes. However, it demands good water quality, a temperature between 80 and 84 degrees, and relatively soft water. Small juveniles are quite colorless, but mature fish develop iridescent colors that change hues as the fish turn in the light. A delicate and subtle beauty, this species needs a little extra care.

Port Cichlid

Cichlasoma bimaculatum: Once placed in *Aequidens*, this species now is interpreted as a species of *Cichlasoma* in the true sense (as opposed to the Central American species placed in that genus but probably belonging elsewhere). The species spawns very much like the Blue Acara, but it gets a little larger, males easily attaining 6 inches

A mouthbrooding species in which both parents take care of the fry, the Jurupari is an eartheater that is tolerant of its tankmates and becomes large enough to deter some of the smaller, more aggressive cichlids.

in length. The species is never striking or colorful, but it is reliable in that it spawns easily and takes excellent care of the young. Only moderately aggressive, this fish is a great first cichlid, along with the Blue Acara, for the budding cichlid keeper.

Pike Cichlid

Crenicichla multispinosa: Although the *Crenicichla* species are deadly ambush predators, many species make good aquarium residents for the special tank, the main requirement being that all members of the tank be too large too swallow. Although the fish are excellent parents,

they aren't overly aggressive as most cichlids go. Once sexual maturity is reached, the males can be told from the females by their larger size (up to 8 inches) and more spangles on the body, but females have better coloration, sporting a bright red belly region. Nearly all of the *Crenicichla* species are cave-spawners, so aquarists use flower pots and PVC piping to accommodate them. Although they are efficient fish predators, they will eat dry and frozen foods. The elongated, often colorful cichlids of this genus are poorly understood and very variable, so many hobbyists choose not to place species names on them. While some species are a foot or more long, others are only 4 to 6 inches. Treat all with the respect given to any predator.

Jurupari

Geophagus (Satanoperca) leucosticta: The Jurupari is an eartheater (taking substrate into the mouth and filtering it through the gills to extract small invertebrates) that has recently been moved from *Geophagus* proper to the genus or group *Satanoperca*. Several almost identical species are confused with it (this one has bright silver spots on the face), but fortunately they are similar in their aquarium needs. These fish are amazingly adaptable to water quality and are found throughout the Amazon, but they have to deal with different water parameters in some coastal areas. A Jurupari can reach nearly a foot in length, but it takes at least three years to do so. It is not aggressive with other cichlids, so its size helps protect it from some of the smaller, more aggressive cichlids. The eggs are laid on the bottom, but two days later the parents chew them to release the larvae, which both parents then guard in their mouths. Both parents take care of the fry.

Green Severum

Heros severus: Green Severums are high-bodied, red-eyed cichlids with many small spots on a greenish or (captive-bred) golden body. They are relatively peaceful biparental substrate-spawners that have long been an aquarium favorite. This Severum attains about 8 to 12 inches in length, not small by any means, but it is relatively mild-tempered. For best results, offer specimens lots of fresh plants as well as live foods. Other almost identical species seem to be in the hobby, but the Green is the only common species.

BIG OR BAD

It takes really dedicated aquarists to keep these fish, as they all get either very large or they are extremely aggressive. Some are both! Certain large species are so aggressive that the hobbyist must put up a grid as a barrier to keep the cichlid from getting at the heater and smashing it to pieces. Cichlids with very aggressive behavior tend to "displace" some of their aggression by attacking objects in the tank. The grid used to protect the heater and other vulnerable equipment is usually the plastic grating used for diffusing big florescent lighting fixtures. Such material can also be used in attempting to breed some of these monsters.

The Severum breeds in the open and is mild mannered except when spawning. A breeding pair should really have a tank to themselves. The male may have more pointed fins than the female and may have dots on the head that the female does not.

Green Terror

"Aequidens" rivulatus: This is one of the species that was excluded from the new definition of *Aequidens*, so quotation marks are placed around the generic name. (To add to the confusion, the true *"Aequidens" rivulatus* may not be the one that is present in the tropical fish hobby.) Green Terrors (which come from Ecuador) reach about 10 inches in length but will spawn at a much smaller size, 5 to 6 inches. The best way to spawn the fish is to obtain six individuals and let them pair up naturally. Once the fish attain their full size, you may need grating to protect the heater and other important fixtures in the tank. A single specimen may be kept as a pet by itself in a tank, as these animals are quite intelligent and learn to recognize their owners and even perform tricks for food. Large specimens are at their best in appearance and coloration, showing bright green edges to the scales and either red or white margins on the tail fin and soft dorsal.

Oscar

Astronotus ocellatus: This species is found all over South America and is one of the most popular of all cichlids in spite of its size. That is partly because the juveniles are so cute and beg for food and attention like puppies. Even larger specimens (to 14 inches and quite high-bodied) are popular because they remain responsive to their keeper. A pair can live together peacefully in a 100-gallon tank. Oscars grow quite rapidly and are quite predatory on any fishes that they can swallow, so they are not suitable for community tanks. They also can be easily bullied by smaller cichlids, as they truly are not aggressive. Oscars have been popular for so long that many artificial color forms have been bred, including a red form that

The Green Terror is aptly named. This is not your average community-tank cichlid.

THE GUIDE TO OWNING CICHLIDS

A very popular fish, the Oscar has been bred in captivity for so long that there now many color forms. They do need a tank that allows them room to move. Those in tight quarters look really miserable.

is quite nice looking and a "red tiger" coloration that isn't bad.

Even though Oscars are specialized fish-eaters, they learn to eat earthworms and dry food in captivity and should have as much variety as possible. Clean water is important for these animals, so at least weekly partial water changes should be made.

Red Terror

"Cichlasoma" festae: Here is another fish that takes a fanatic to keep it, but it brings enough attractive qualities to inspire such fanaticism. A specimen in good condition has stunning coloration, and its behavior is not only fascinating but downright spectacular. A single specimen can be kept in a community tank of large and rough cichlids, but the tank should be quite large—of at least 200 gallons capacity. Spawning is best done when they are relatively small, as pairing at full size is quite tricky. Full adults are easily 12 (females) to 18 (males) inches in length and are quite formidable. They will spawn at a size of less than 6 inches. A full-grown compatible pair with young is a sight to behold, and even the aquarist is tolerated only within narrow parameters. Red Terrors are boldly marked with black and red vertical bands in juveniles, holding this pattern into adulthood in females; males may become largely iridescent green on the sides and develop wrinkled lips.

Cichlids of Africa

WEST AFRICAN CICHLIDS

The many cichlids of western Africa are no longer as popular as they were 50 years ago, being to a great extent replaced by the species from Lakes Malawi and Tanganyika, but they still have their fans. Especially popular are several species known as kribs, which usually are small (under 5 inches), long-finned, delicately colored, and moderately

Chromidotilapia guentheri is a mouthbrooding species in which the male incubates the eggs while the female stands guard.

Jewel Cichlids are substrate spawners with both parents caring for the fry. There are several cichlids of this type that vary somewhat in appearance. All are noted for their fierce protection of their young.

peaceful as well as good spawners; female kribs usually have bright pink or red on the belly when ready to spawn. Some of these fishes are among the most familiar beginner's cichlids.

Guenther's Cichlid

Chromidotilapia guentheri: There is no popular name for this species, but it certainly has become popular with advanced cichlid hobbyists. One of the reasons is its very interesting behavior. This is a mouthbrooding cichlid in which the male incubates the eggs but the female does not abandon him: she guards him. In the meantime, she is able to feed to help build up energy reserves for the next clutch of eggs. She also helps guard the free-swimming fry once he releases them. It is quite possible that this species is naturally monogamous in the wild. The species was once known as *Pelmatochromis guentheri.*

This 6- to 7-inch kribensis-type cichlid is found in coastal western Africa from the Ivory Coast to Cameroon and is exceptionally hardy.

Common Jewel Cichlid

Hemichromis guttatus: Actually, there are several jewel cichlids that look quite similar, but they vary somewhat in size and intensity and detail of coloration. This seems to be the proper name of the most common species in the hobby.

Jewel cichlids are substrate-spawners noted for their fierce protection of their young and adaptability to a variety of water conditions. They are one of the toughest and most aggressive cichlids. This species, which has a large black spot near the center of the side, a large spot on the operculum, and many rows of small silvery blue spots on the face and sides, all against a brilliant red color, is from western Africa and reaches 4 inches in length, females smaller than males. Jewels are noted as among the best and most faithful cichlid parents.

Krib

Pelvicachromis pulcher: Known through most of its history in the aquarium hobby as *Pelmatochromis kribensis,* this is one of the most popular dwarf African cichlids (adults 4 inches or less) and one that generally does better and looks its best in soft water. It is peaceful for a cichlid, but it is too aggressive to be kept with small South American species. Like other related species, the female has a bright red blotch on the belly. She may place up to a hundred eggs on the roof of a cave and not allow the male to re-enter after he fertilizes them. The young should be kept at a nearly neutral pH if you wish both sexes to develop.

Tilapia buttikoferi.

the species are quite attractive and long have been kept in aquaria. Originally all were in the genus *Tilapia,* but today several distinct groups are recognized. These include *Oreochromis,* which are mouth-brooding cichlids, with the female brooding the eggs; *Sarotherodon,* which is intermediate between *Oreochromis* and *Tilapia* in being mouthbrooders, but the males and females stay together and either the male or female or both may pick up the eggs once they have been fertilized; and the true *Tilapia* species, which are substrate-spawners with both parents tending the young. Some popular aquarium species are *Oreochromis mossambicus, Sarotherodon melanotheron,* and *Tilapia zillii.* Although each gets a little large (often to 10 inches), they spawn readily and have a handsome appearance. They are among the hardiest and most adaptable cichlids, surviving and reproducing in all but extreme temperature and pH conditions.

Kribs should be kept in a well planted tank with lots of hiding places.

Tilapias

These fishes are important because they have been too successful, escaping from ponds after having been introduced as food fishes around the world. Some of

CICHLIDS OF LAKE VICTORIA

The cichlids of Lake Victoria, a very shallow Great Lake at the northern end of the chain, have been of particular interest lately because they are all endangered. Not only did many species disappear after the 6-foot Nile Perch, *Lates niloticus*, was introduced to the Lake as a food fish, but forest destruction and farming have made for muddy runoffs and great increases in silt and thus loss of clarity in the Lake. Many of the species discussed in the hobby literature 30 years ago appear to have become extinct, but newly found (and perhaps evolved) species have to some extent replaced them. To many hobbyists and some scientists the Victorian cichlids are uninteresting as all look much alike, apparently being derived from a single ancestor relatively recently. Almost all have been treated as the genus *Haplochromis*, but recently this has been broken into numerous poorly defined genera. Similar species live in rivers entering the Lake.

Unfortunately, not many species of the Victorian cichlid flock (estimated at some 600 to 1000 species, living and recently extinct) have made it to this country. All the aquarium species are maternal mouthbrooders that should be kept in groups of one male to several females. This can be difficult because shippers tend to send more males than females, as males typically are much more colorful than females. Lake Victoria cichlids vary from tiny (an inch or less) silvery species to large (8 to 12 inches) species with dark vertical bands and brilliant colors on the sides and face (usually including a dark stripe down from the eye). They may be planktonic feeders in open water or feeders on invertebrates and fishes in rocks or sand. At the moment, it seems virtually impossible to place names on any species shipped from the Lake or even to be certain that different individuals belong to the same species. An exception in the colorful 4-inch *Pundamilia nyererei*, in which males are bright reddish brown on the back with golden lower sides bearing many wide vertical blue-black bars; females are tan but have a similar pattern.

LAKE TANGANYIKA CICHLIDS

Unlike the cichlids of Lake Victoria and Lake Malawi, the cichlids of Lake Tanganyika present an incredible diversity. They vary in their methods of reproduction and also vary greatly in body shape. Many of the cichlids of Tanganyika are small, a convenient size for aquaria, so it is possible to have a large community tank in the living room with an elegant *Tropheus*, some snail shells on the

Nyereri is an exceptional fish for a Victorian in that it can be readily identified.

bottom occupied by a pair of shell-dwelling cichlids, some eye-popping *Lamprologus* (*Neolamprologus*) *leleupi*, stately *Lampro-*

logus (Altolamprologus) calvus, and exotic *Cyphotilapia frontosa*. The many different shell-dwelling cichlids are a show in themselves. Tiny fellows that are barely an inch long, they ferociously defend their shells, chasing after much larger fishes. *Tropheus* species are among the most elegant of fishes in appearance and bearing, but only one can be kept to a tank without ferocious fighting among them. The only alternative is to keep at least ten of them in a tank. For some reason, that is about the magic number to diffuse the aggression among them. One of the most beautiful of tanks involves combining some species from Lakes Malawi and Tanganyika. When it is done, the species must be selected not just on looks but on levels of aggression. Generally speaking, mbuna (Malawi) can be combined with *Tropheus*, while *Aulonocara* species (Malawi) can be combined with most of the *Lamprologus* types.

Most of the familiar cichlids from Lake Tanganyika (those related to *Lamprologus*) are substrate-spawners: they lay their eggs on a rock, in a cave, or in a pit and protect them and the fry after they have hatched. However, there also are many mouthbrooders in the Lake. In most mouthbrooders from the Lake, the male digs a nest pit and stands near it, inviting ready females in to spawn. A female comes in, lays a few eggs, and immediately picks them up in her mouth, sometimes before the male has fertilized them; in that case, they will actually be fertilized within her mouth as she nuzzles the base of his anal fin. The fry and young are cared for by the female, while the male continues to indulge in maintaining his nest and mating with other females.

Brichardi

Chalinochromis brichardi: Most people believe that there are many species of *Chalinochromis*, as there is much variation in members of the genus. Hobbyists tend to refer to the specimens with the lines down the body as "*bifrenatus*," a name actually better associated with the genus *Telmatochromis*. The ones referred to as *brichardi* in the hobby are those with a beige body and a mask around the head. These cichlids look benign and are slow-moving, gliding along the rockwork, but they can be quite aggressive. The best way to spawn them is to get at least six young and raise them together. They spawn in caves or under rocks, laying just a few eggs

Cyathopharynx furcifer. Furcifers are among the most desirable of the African cichlids.

THE GUIDE TO OWNING CICHLIDS

Chalinochromis brichardi.

at a time. You may not know young are present until fry start appearing from the rocks. Young do well on dry food and the incidental food that they are able to pluck from the rocks, but they will prosper on newly hatched brine shrimp. The species attains a length of about 4 inches, with rare individuals attaining a length of 6 inches.

Furcifer

Cyathopharynx furcifer: This is one of those species that may actually comprise several different species, as there is much variation according to the locality in Lake Tanganyika. One of the highly prized species, it is a quite showy fish reaching 8 inches in length. The males pick the highest point upon which to build a nest, carrying up sand. The females tend to pick the male with the highest nest. Males tend to be blue to purple in color and fairly shimmer as they display before the females. Females brood the eggs and fry in their mouths.

This species feeds upon invertebrates in the water column, but it will also take dry foods as well as frozen formulas. It also thrives on newly hatched and adult brine shrimp. The fry normally are raised with newly-hatched brine shrimp as a first food, but they will also take fine powdered food.

Frontosa

Cyphotilapia frontosa: This cichlid lives in the perpetual twilight of the Lake at depths of about 100 feet and feeds upon invertebrates and small fishes. At those depths the supply of plankton is greatest in the daytime, so many fishes feed on the plankton, and *C. frontosa* feeds upon them. This is a very popular fish that reaches a length of nearly 14 inches. One of the reasons for its popularity, in addition to an exotic appearance (vertical blue-black and silver stripes, the male with a large hump on the head), is that it is an easy-going, fairly sedentary fish that does well in the aquarium, usually just

Cyphotilapia frontosa **is a deep-water fish that nevertheless does quite well in the aquarium.**

hanging in the water near the bottom. When it does move, it does so without apparent effort. Be aware that it will gradually eliminate any smaller fishes from the aquarium.

The male constructs a rudimentary nest, and the female lays up to 20 large eggs. She broods these in her mouth for nearly a month, releasing young that are nearly an inch in length and look very much like miniature adults. Males don't show the hump on the head until they reach about 6 inches in length. Even hobbyists who aren't fond of this fish breed it because it is always valuable commercially.

Cyprichromis microlepidotus: This is my nominee for the most uncichlid-like fish in the Lake. It is a plankton-feeder that to the untrained eye looks more like an Australian rainbowfish (*Melanotaenia*) than a cichlid. There are many different color variations of this one species. The female and male go through a mating ritual above the bottom, the female circling the male before dropping eggs. She picks up the eggs after they are

fertilizes and incubates them in her mouth. The young are guarded by the mother. Males defend a territory above the sand.

Julie

Julidochromis ornatus: There are several species of *Julidochromis*, including the very popular *Julidochromis marlieri* and the small *Julidochromis transcriptus*. They are all quite similar in appearance and behavior. This species attains a length of about 3 inches. Although exceedingly quiet fish, they can be aggressive, but they have a distinctive personality that has made them quite popular. For one thing, they orient themselves to rocks. Wherever the rock is means "down" to them. They will glide up the side of a rock, skim along its surface, sweep down the other side, and then go underneath upside down. They nearly always have a home in a spot underneath a rock. Whenever they are in their home, they will be upside down. Even when feeding at the surface of the water, they are often upside down, orienting themselves with the surface.

Julidochromis ornatus. **Julies orient themselves to the rocks they call home.**

The best way to breed them is to get about six young; eventually a pair will form and drive the others off. They tend to bond as a pair for life, raising only a few young at a time. The eggs are jade colored and are laid on the roof of a cave. You probably won't know the fish have spawned until young begin gliding along the rocks, much like adults. Both parents defend the eggs and the young. Adults feed upon invertebrates that they find in the biocover of the rocks, but they eat nearly anything in captivity. The fry do well on powdered foods, but they do even better on newly hatched brine shrimp.

Brevis

Lamprologus (Neolamprologus) brevis: This species is a shell-dweller (in old snail shells) that barely reaches 2 inches in length. The female is even smaller. An aggressive little fish, it survives in sandy areas where large fish prowl, living in the safety of the shells. Males hoard snail shells, as that way they get to have more females, one female for each shell. The female lays her eggs in the shell and the male fertilizes them there. Sometimes he can't fit in the shell, but his anal fin helps guide the milt into the shell. The fish feed upon invertebrates that they find in the water and in the sand. In the aquarium they eat anything, including dry food.

Brichardi

Lamprologus (Neolamprologus) brichardi: One of the most ubiquitous cichlids in the lake is *Lamprologus (Neolamprologus) brichardi*. It feeds upon the

Lamprologus brevis is a shell-dweller that will take up residence in escargot shells.

plankton in the water, and the dentition, which is quite impressive, may have evolved for defense as well as for plucking invertebrates from rocks. A secret of its success is that family members help defend the young. It is touching and amusing to see tiny fish not quite an inch long defending younger siblings. Although frilly in appearance, this cichlid is quite formidable, which helps account for its success in the lake and makes it a species that can be combined with many other species once it has established its breeding colony. Although this fish spawns among the rocks and in caves, it normally places the eggs on the substrate, rather than upon the roof of a cave. Only a few eggs are laid at a time, but the survival rate is high. The only drawback to this great little species is that a pair can eventually end up taking over a tank as its family increases in size, but it is a simple matter to harvest some young for sale. So successful has this type of fish been that it is found throughout the Lake, and it is difficult to know how many real species it includes. In any case, this is a popular and elegant fish.

Calvus

Lamprologus (Altolamprologus) calvus: While most Lake *Lamprologus* species might be called *Neolamprologus* by some modern hobbyists and ichthyologists, this species would be put into the subgenus or genus *Altolamprologus*. It has a very compressed body that allows it to fit into crevices in rocks and go after invertebrates and the young of other fishes. The large, oblique mouth makes it look neandertal, but it is functional for getting its prey. Surprisingly, this compressed species likes to use snail shells for spawning, too, but it is less demanding than some of the other species that wouldn't even think about spawning without a shell. A cave will do fine, but the female tends the eggs while the male stands guard outside.

Leleupi

Lamprologus (Neolamprologus) leleupi: This colorful species is generally found in deep water. Though reaching a length of only 2 or 3 inches, it is extremely aggressive, especially with its own kind, but it looks for trouble with everyone. It will breed in shells but settles for caves. Most breeders keep them in a community tank of sorts for spawning, putting about six *leleupi* in a tank of mixed Tanganyikan cichlids. Ceramic caves are often used, with holes or slits just big enough for the female and male to enter. The breeder knows that there is a spawning when the female consistently stays within the cave and the male guards it. A standard practice is for the hobbyist to place his or her thumb over the opening and remove the cave to a rearing tank. This usually is done when the young start to stick their noses out of the cave. The female can be netted out of the rearing tank at the hobbyist's convenience. Feed the young with newly hatched brine shrimp and colorfoods. If they don't get it when they are young, they won't retain the orange or yellow coloration.

Ocellatus

Lamprologus (Neolamprologus) ocellatus: This shell-dweller reaches about 2 inches in total length in the male, the

Lamprologus ocellatus **moves a bit of sand. Many cichlids are quite industrious about arranging their tanks.**

female smaller. Both sexes have a dark spot on the gill cover, but it is more pronounced on the males. Although both sexes can enter the snail shell, the male spends more time outside in a guard mode in fierce defense of his lair, and he has sharp canine teeth.

Lamprologus (Neolamprologus) sexfasciatus: This is one of the more popular Lake species, although it is often confused with *Lamprologus tretocephalus*. (The latter species has only five bars, while this

Lamprologus sexfasciatus **is found in the southern part of Lake Tanganyika.**

one has six.) Both species seem to specialize in preying upon snails, but they thrive on prepared foods of all type in captivity. There are several color forms in the Lake, but the most common in the aquarium have been the type with a blue-gray background and one with a yellowish background. This species normally reaches about 3 inches but can attain 5 inches in time; it will spawn at a length of only 2 inches. Like so many lamps, this one likes to spawn in caves, and it likes lots of rockwork.

Fourspine Cichlid

Lamprologus (Neolamprologus) tetracanthus: This species can easily reach 8 inches in length, and it is a rough cichlid that doesn't really need a lot of rocks to feel secure. A pair will build a nest in the sand, although they will utilize sandy

The Fourspine Cichlid eats snails by sucking the snail out of the shell.

areas under rock ledges when given the choice. The species is one of the many snail-eaters, although it is a generalized predator and will take fish species, too, including some of the small cichlids that live in snail shells!

Tret

Lamprologus (Neolamprologus) tretocephalus: This is one of the cichlids for which there is always a commercial demand. It is a nice size, reaching about 4 inches but spawning at 2, and it has nice coloration. Both parents herd the young out in the open and give them vigilant attention and defense. This makes the species a contrast to so many of the species that spawn in caves, shells, and under rocks. They are more like New World cichlids in that they are more obvious when they spawn. Of course, in a community tank they can make life quite hard on the other species when they spawn. *Lamprologus tretocephalus* feeds on snails in the Lake (as well as fishes), but it takes a much wider diet in the aquarium.

Boops

Ophthalmotilapia boops: This large-eyed cichlid is a maternal mouthbrooder with an interesting breeding pattern. The males construct nests and try to entice females to them. The female lays golden eggs. When she picks them up, she mouths the ends of the male's ventral fins, which triggers the male to release sperm, fertilizing the eggs in her mouth. The gold knobs at the ends of the male's fins resemble the eggs in color and shape. The

Boops is a plankton feeder in the Lake.

shimmering colors of the male have made this fish an elite species among the enthusiasts of Lake Tanganyikan cichlids, but it is relatively aggressive and large (over 6 inches).

Polyodon

Petrochromis polyodon: All the fish in this genus have so many teeth that their lips protrude; the teeth are used to scrape algae off rocks. Although beautiful and available in many color varieties, including crimson, this species can be a problem to keep because of the pugnacity of the individuals with one another. The species is a maternal mouthbrooder and reaches about 6 inches in length. Some of the other species in the genus reach a length of nearly 2 feet.

Polyodon is a tough fish that reaches a respectable 6-inch size.

Duboisi

Tropheus duboisi: There are over a hundred known different color varieties of *Tropheus*, but only a few of them have been described as separate species. In *Tropheus duboisi* the juveniles have a different color pattern from the adults, being black and covered with numerous white spots. Adults have a narrow yellow or red band on a black body. All *Tropheus* are maternal mouthbrooders, with the females carrying the few but very large eggs for up to a month. The mother protects the large young for a few days after release.

Tropheus feed together in groups, browsing upon the algae-covered rocks of the Lake. Though herbivores, they can be quite aggressive with one another in aquaria. One of the secrets to keeping them is to keep large numbers of a species. It is interesting to note that different color variants can be kept together in a tank, and they don't seem to cross-breed. Do not feed much live food, as it can be deadly to them, especially if they get a steady diet of such fare. Feed dry food that has a lot of plant matter in it. *Tropheus* should have the excellent water quality and high pH that is required by all the other cichlids of Lake Tanganyika.

CICHLIDS OF LAKE MALAWI

The only drawback to the cichlids of Lake Malawi is that they are all mouthbrooders, but there is much variation to the method employed. There also is much variation in feeding behavior,

Duboisi are the best-known of the many forms of *Tropheus.*

and the colors and movements of many species are so dazzling that this group is probably the one most responsible for the popularity of the cichlid family with tropical fish hobbyists. As for scientists, they long have studied both behavior and classification of these cichlids, and they have found both areas of study a delight of complexity. Currently 300 to 400 species are recognized in the Lake, with at least as many more known but not yet described. Dozens of species, especially of the colorful rock-dwelling forms known as mbuna, are imported and find a place in the tanks of dedicated hobbyists. A Lake Malawi aquarium should have hard, alkaline water and generally considerable rockwork on the bottom. Many Malawi cichlids are very aggressive with others of their species, so some hobbyists keep several species in an aquarium and give them plenty of areas to establish territories around caves and rocks. As a rule, males are much more colorful than

females in mbuna, but in the open-water forms (which may be brilliant blue) the sexes are more equal in coloration. The higher-bodied open-water species formerly were mostly in the genus *Haplochromis* (but now are placed in many genera), leading to the term "hap" as part of the common name of many.

Because so many species are imported from Lake Malawi and kept by hobbyists, only a few of the most popular species can be mentioned here. There is no doubt that the cichlids of Lake Malawi are among the most beautiful fishes in the world, and they have done a lot for the popularity of the cichlid hobby in the last 30 years. Some cichlid specialists become disdainful of these fishes because so many beginners keep them, but I am always learning something new about Malawi cichlids, and so are scientists.

Blue Peacock

Aulonocara hansbaenschi: This is one of the peacock cichlids that come in colors ranging from flaming yellow to bright blue on the head and fins. In this species the males are brilliant blue with a white edge to the dorsal fin and just a touch of rust on the chest and ventral fins. More quiescent than many other species in the Lake, *Aulonocara* are generally deepwater species with special sensory pores on the jaws that enable them to detect crustaceans in the sand. They hover over the sand, "hear" a shrimp, and then suddenly dive in and grab a meal. The best way to breed these guys is to keep one male with several females, and that is the best

Peacocks are the aristocracy of African cichlids.

method with most of the cichlids in this Lake. Females often are almost as colorful as males.

Trewavas

Labeotropheus trewavasae: This is one of the mbuna that fueled the original mania for African cichlids in the 1970s, as it comes in many different color forms. The most pretty (in my view) has a bright blue body with an orange dorsal fin. Both sexes display these colors. A mouthbrooder like all the other Malawi cichlids, spawning is rarely observed as the male does not make a nest and spawnings take place quickly. The female incubates a few large eggs in her mouth for several weeks and then releases young that are miniatures of the adults. The mouth is especially adapted for scraping algae off the rocks and is located under a blunt nose that often projects considerably.

Auratus

Melanochromis auratus: This is another species that helped spark the boom in cichlid interest among tropical fish hobbyists. It was first known as *Pseudotropheus auratus* back when Lake Malawi was known as Lake Nyassa. Spawning is very similar to the foregoing

species; however, the male changes coloration upon developing sexual maturity. Juveniles and females are bright yellow with two dark brown lines on the side and one in the dorsal fin, but males become deep blue-black on the belly and upper side, with a silvery stripe from the eye onto the tail; the back in males may remain yellowish brown. Like most mbuna, it is a rock-dweller and seldom reaches 4 inches in length.

Livingstonii

Nimbochromis livingstonii: This species is too large (10 inches) for most hobbyists, but I couldn't resist listing it because of its unusual behavior. It lies on its side and plays "possum," waiting for little fish to pick at the corpse. When they get close enough, the corpse comes to life and grabs a meal. It is a deep-bodied *Haplochromis*-type fish that commonly is whitish tan with large brown spots partially fused into vertical and horizontal bars.

Electric Blue Hap

Placidochromis electra: This slab-sided cichlid from deeper waters of Lake Malawi is noted for the extremely

Livingstonii is a trickster that "plays possum" to get its meal.

THE GUIDE TO OWNING CICHLIDS

The Electric Blue Hap is a flashy fish that can easily be kept in quantity in the aquarium due to its rather placid nature.

brilliant blue tones of the male, which also carries a nearly black bar obliquely behind the head. Though it reaches 6 inches in length, it has a small mouth and is relatively nonaggressive, often being kept in groups of up to a dozen fish in very large aquaria.

Tropheops

Pseudotropheus tropheops: This is another mbuna that came in with the first Malawi shipments and led to Malawi hysteria. It is yet one more cichlid that comes in many guises, and one of my favorite color varieties is one in which the male is a brilliant blue and the female is a bright orange. Scientists are going crazy trying to sort out the genus *Pseudotropheus* (now often split to include this species in the genus *Tropheops*). This species and its close relatives have a short, rounded snout and relatively large eye as well as a somewhat deeper body than most mbuna.

Zebra

Pseudotropheus zebra: No discussion of Lake Malawi cichlids could ignore this species (actually a species group), long the most popular of all Malawi fishes. This mbuna was one of the earliest importations and become noted for its aggressive males of brilliant silvery blue with many wide black vertical bars. The mouth is somewhat oblique. Like other mbuna (and most other Malawi cichlids), the anal fin of the male bears large golden spots outlined in black, so-called egg dummies that the female mouths after

The Zebra is the quintessential mbuna from Lake Malawi. This is the fish that brought African cichlids into the forefront of the aquarium hobby.

taking her eggs in her mouth; at the same time, the male is releasing sperm, so the eggs are fertilized in her mouth. Scientists have found egg dummies endlessly fascinating. Currently some scientists place this species and its many similar relatives in a distinct genus variously known as *Maylandia* or *Metriaclima*.

This short coverage of the cichlids should just serve to whet your appetite, as there are dozens of equally colorful, interesting, and keepable species in the family. Once you try a few species, you are sure to be hooked on cichlids.

Index

Photo Credits

Bob Allen, 10, 16, 54, 58, 61, 63
Dr. Herbert R. Axelrod, 41
Bede Verlag, 30
C. Chou, 15
Don Conkel, 40
Dr. Harry Grier, courtesy of FTFFA, 48
Ad Konings, 60 (bottom)
Horst Linke, 13, 42, 43 (bottom)
Oliver Lucanus, 45 (top), 56
Brian Middleton, 26
Aaron Norman, 37, 39, 47, 53, 59
John O'Malley, 25 (top, bottom)

Jiri Palicka, 10
MP. & C. Piednoir, 1, 11, 14, 19, 20, 22, 27, 28, 32, 35, 37,
    40, 44, 46, 50, 51, 52, 55, 62, 63
Harry Piken, 49
Hans-Joachim Richter, 4, 8, 12, 18, 36, 38 (bottom), 57
Andre Roth, 43
Dr. Jurgen Schmidt, 7
Vince Serbin, 24
Mark Smith, 38 (top), 60 (top), 62
Wolfgang Sommer, 52 (bottom)
Alf Stalsberg, 33
Edward Taylor, 56, 59